T0073449

Studies in Fuzziness and Soft Computing   277

Editor-in-Chief

Prof. Janusz Kacprzyk
Systems Research Institute
Polish Academy of Sciences
ul. Newelska 6
01-447 Warsaw
Poland
E-mail: kacprzyk@ibspan.waw.pl

For further volumes:
http://www.springer.com/series/2941

Lotfi A. Zadeh

Computing with Words

Principal Concepts and Ideas

 Springer

Author
Prof. Lotfi A. Zadeh
University of California
Berkeley
USA

ISSN 1434-9922   e-ISSN 1860-0808
ISBN 978-3-642-27472-5   e-ISBN 978-3-642-27473-2
DOI 10.1007/978-3-642-27473-2
Springer Heidelberg New York Dordrecht London

Library of Congress Control Number: 2012937218

Printed on acid-free paper

Springer is part of Springer Science+Business Media (www.springer.com)

Dedicated to Janusz Kacprzyk, Jerry Mendel,
Behzad Kamgar-Parsi, Enric Trillas and Paul Wang
in appreciation of their support for Computing with Words

Foreword

It is a great honor and pleasure for me to write this foreword to this very special and unorthodox book project on computing with words (CWW, for short) by Professor Lotfi A. Zadeh. This book is a logical sequence of slides on CWW which have been presented for years by the author at a multitude of conferences, congresses and other scientific gatherings.

The first question a reader can ask is that it may be somehow unclear how and why a book can be comprised of slides and not, as usually, of a plain text, possibly with some formulas, tables, figures, etc. as it is usually the case. This question is however related to a more fundamental question, why plenary talks are needed if one could be able to read the contents in the form of an article. There is clearly no simple answer, and yet people all over the world accept plenary talks as a vital part of conferences, both small and large. In plenary talks authors can convey in a more general, and hence more inspiring way, new ideas and views, can reach a wider audience, possibly from beyond their areas of interest and activity. From the slides, moreover, the reader can probably readily see the author's opinion and views regarding various issues.

The author of this book, Professor Lotfi A. Zadeh, is a special person in science and many other realms. Though to many younger people he is best known as the founder of fuzzy sets theory, one should bear in mind that before his development of the theory of fuzzy sets he has been one of the most inspiring and prominent representative of fields like systems science, automatic control, signal processing, and even optimization. Just to give an example, he has been one of the originators of the famous state space approach that has been for decades used by thousands of people in hundreds of areas in all kinds of mathematical modeling, optimization, control, etc.

After such an illustrious career in "hard" and "crisp" areas of research, he has been – probably since the early 1960s – coming to a conclusion that to be able to realistically model real world, usually human centric systems, "softer" tools and techniques are needed, and he proposed in the mid-1960s the theory of fuzzy sets. This was a remarkable achievement in that, for the first time, a conceptually obvious and intuitively appealing set of tools and techniques was proposed to handle imprecise information exemplified by "large number", "high temperature," "much more than," etc. The use of natural language, the only fully natural means of articulation and communication for human beings is clearly a source of that uncertainty. Fuzzy set theory has since its inception enjoyed a tremendous, unprecedented growth, and it is difficult to find another area in which in such a short time so many relevant papers have been published, so many people have been working in, so many citations have appeared, etc.

The success of fuzzy sets theory, which after a decade or two has become a well established academic discipline with thousands of papers and hundreds of active scholars and researchers, has not stopped Zadeh from further work, and his inspiring role for the world scientific community has not diminished. His ideas which have influenced many people, triggered new research and scholarly directions are too numerous to mention.

One of them is *computing with words* (CWW), termed sometimes also in a longer form as *computing with words and perceptions* (CWP), originated by Zadeh in the early or mid-1990s, which has rapidly attracted much attention from scholars and researchers from around the world. I was privileged to be part of those who have been since the very beginning convinced that CWW is a real breakthrough and will have a lasting impact on many areas in which systems and processes should be modeled, and we are to operate in a human centric context in the sense that human specific features are crucial, among them the human willingness and propensity to use natural language. As a result, a two volume book on CWW edited by Zadeh and myself was edited in 1990 and appeared in the present book series. This book has helped to establish the new area of CWW, attract many new people working in both theory and applications, and has also–in my case – focused my research interests and activity for years.

What is exactly CWW, will become clear to the reader while reading this book, maybe – better to say – by following the line of thought reflected in the slides. Moreover, he or she will be able to follow the evolution of the concept which has clearly culminated in what might be called CWW^2, the second generation of CWW. This has been for some time advocated by Zadeh who has once again seen challenges and opportunities, and has found ingenious ways to formulate and implement them.

Basically, as the constituent words of CWW clearly indicate, there is "computing (computation)" and "words (language)." It deals therefore with a natural problem: since for human beings the only fully natural means of articulation and communication, and since in science we are in general trying to perform a quantitative analysis, then it would be beneficial to develop a computational system in which the objects of computation would be words, phrases and propositions instead of numbers. One can obviously notice that a similar problem is addressed by various tools of computational linguistic, natural language processing, generation, etc. These areas, however, are not the same; they have different agendas and use different paradigms. From our point of view it is enough to say that they fail to grasp and process imprecision which is a crucial characteristic feature of natural language. They also fail to adequately and fully represent computational problems stated in natural language. What is so common, at least in a vast majority of applied science, is that there is a real problem—a problem which is perceived by human beings who use natural language to describe it, and then—in traditional approaches—are forced to jump from natural language to numbers because the tools available do require that. CWW can do this, i.e. provide a direct transition from natural language to computation, both at the level of representation and processing! I think that this feature alone is enough for attracting a large community of readers to the fascinating area of CWW.

I will not write more about the fascinating field of CWW, various rationales and solutions employed, and results obtained because they will be shown in this book. My main message is that the richness of ideas and vision of the author will certainly imply that many people who have either been not exposed to CWW or have not been fully convinced about its power, will appreciate it and try to look deeper into its foundations or apply it to solve their problems.

I am sure that the fully innovative idea of this book as a collection of slides will be found convincing and inspiring for many individuals from many fields of science and technology and will trigger further research in the fascinating field of computing with words. As the Editor in chief of Studies in Fuzziness and Soft Computing, the largest book series on fuzzy sets and soft computing in the world, I am honored and proud that this innovative Zadeh's book will appear therein.

January 2012   Janusz Kacprzyk
Editor in chief
Studies in Fuzziness and Soft Computing

Preface

I have been urged by some of my close friends to write a book on Computing with Words. At this stage of my life, it is difficult for me to do so. I am grateful to Springer and to my good friend, Professor Janusz Kacprzyk, Editor of the series on "Studies in Fuzziness and Soft Computing," for making it possible to describe in print the principal concepts and ideas which underlie Computing with Words.

I consider Computing with Words to be an important chapter in my work. What lies between the covers of this volume is a substantially revised and updated version of a PPT file of a lecture which I gave at USC in 2010.

What is Computing With Words (CWW) and what does it have to offer? I will preface my response with a bit of history.

In 1996, I wrote a paper entitled "Fuzzy Logic = Computing with Words," published in the IEEE Transactions on Fuzzy Systems. (Zadeh 1996) What I meant by equality was that, in large measure, applications of fuzzy logic involve the concept of a linguistic variable. (Zadeh 1973, 1975a) My paper evoked a number of critical comments, suggesting that Computing with Words is merely another name for Natural Language Processing.

This was, and remains, a common misconception. The concepts and ideas which were described in my 1996 paper have little in common with Natural Language Processing. Computing with Words and Natural Language Processing have different agendas and different application areas.

Basically, Computing with Words is a system of computation in which the objects of computation are words, phrases, propositions, questions, commands and other types of semantic entities drawn from natural language. What is important to note is that there are two levels of complexity in Computing with Words.

In Level 1 CWW (CWW1), the objects of computation are words, phrases and simple propositions such as Robert is tall, X is small, if X is small then Y is large, etc. Today, most applications of fuzzy logic are applications of Level 1 CWW. The same applies, more broadly, to research literature.

At this juncture, there are over 250,000 papers and books with fuzzy in the title. What should be underscored is that, in large measure, this literature relates to Level 1 CWW.

In Level 2 CWW (CWW2), the objects of computation include possibly complex propositions exemplified by: Usually it takes Robert about an hour to get home from work. It is very unlikely that there will be a significant decrease in the price of oil in the near future. Brian is much taller than most of his close friends, etc. In Level 1 CWW, there is no need for semantics of natural languages. In Level 2 CWW, semantics of natural languages plays an important role.

The genesis of Level 2 CWW was my 1999 paper, "From computing with numbers to computing with words—from manipulation of measurements to manipulation of perceptions," published in IEEE Transactions on Circuits and Systems 45. (Zadeh 1999) In this basic paper, a foundation was laid for what I called the computational theory of perceptions, or CTP for short.

The point of departure in this theory is the observation that a natural language is basically a system for describing perceptions. In CTP, this observation suggests a key idea—the idea that computation with perceptions can be reduced to computation with natural language descriptions of perceptions. What is used for this purpose is the formalism of Level 2 CWW.

The centerpiece of Level 2 CWW is what I call restriction-based semantics of natural languages, or RS for short. In Level 2 CWW, RS is used in the main for precisiation of meaning. Furthermore, in RS, an unconventional definition of a proposition is employed.

More concretely, RS serves to translate a natural language into a mathematical language in which the objects of computation are well-defined—though not conventional—mathematical constructs.

For this purpose, traditional approaches to representation of meaning of propositions are put aside. Instead, a proposition, p, is defined as a restriction (Zadeh 1975b) on the values which a variable X—a variable which is associated with p—can take. The concept of a restriction plays a key role in restriction-based semantics.

Restriction on p is represented as X isr R, where X is the restricted variable, R is the restricting relation and r is an indexical variable which defines the way in which R restricts X. Restriction-based semantics is the only system of computation which has the capability to precisiate natural language. Restriction-based semantics employs concepts and techniques drawn from fuzzy logic.

Computing with Words has an important ramification for mathematics. It opens the door to construction of mathematical solutions of computational problems which are stated in a natural language. Traditional mathematics does not have this capability. The need for this capability will become increasingly evident as we move further into the age of automation of everyday reasoning and decision-making.

Concluding Remark. In conclusion, I believe that in coming years, Computing with Words will gain recognition both as an important tool for the conception, design and utilization of intelligent systems, and as a system of computation which empowers mathematics to construct mathematical solutions of computational problems which are stated in a natural language.

Important Note. What should be underscored is that what follows is restricted to exposition of only Level 2 CWW, since Level 1 CWW is amply covered in the literature.

February 2012   Lotfi A. Zadeh

Table of Contents

INTRODUCTION

WHAT IS COMPUTING WITH WORDS (CWW)?

- There are many misconceptions about what Computing with Words is and what it has to offer. A common misconception is that CWW and NLP (Natural Language Processing) are closely related. In fact, this is not the case. CWW and NLP have different agendas and address different problems. A very simple example of a problem in CWW is the following.

L.A. Zadeh: Computing with Words, STUDFUZZ 277, p. 3–37
springerlink.com

EXAMPLE

CWN

- Dana is 25 years old
- Tandy is 3 years older than Dana

- Tandy is (25+3) years old

CWW

- Dana is young
- Tandy is a few years older than Dana

- Tandy is (young + few) years old

- In CWW, young and few are interpreted as labels of fuzzy numbers. Fuzzy arithmetic is used to find the sum of young and few.

ESSENCE OF CWW

- In essence, CWW is a system of computation in which the objects of computation are words, phrases and propositions drawn from a natural language. The carriers of information are propositions. It is important to note that CWW is the only system of computation which offers a capability to compute with information described in a natural language.

KEY POINT

- *In particular, CWW opens the door to construction of mathematical solutions of computational problems which are stated in a natural language. For convenience, such problems will be referred to as CNL problems. Here are a few simple examples of CNL problems.*

- *(a) Most Swedes are tall. What is the average height of Swedes? (Solution in Phase 2—Computation)*

6/263

- *(b) Probably John is tall. What is the probability that John is short? What is the probability that John is very short? What is the probability that John is not very tall? (Solution in Appendix)*

- *(c) Usually, most United flights from San Francisco leave on time. I am scheduled to take a United flight from San Francisco. What is the probability that my flight will be delayed?*

7/263

- *(d) Usually Robert leaves his office at about 5 pm. Usually it takes Robert about an hour to get home from work. At what time does John get home?*

8/263

- *(e) X is a real-valued random variable. Usually X is much larger than approximately a. Usually X is much smaller than approximately b. What is the probability that X is approximately c, where c is a number between a and b?*

9/263

- *(f) A and B are boxes, each containing 20 balls of various sizes. Most of the balls in A are large, a few are medium and a few are small. Most of the balls in B are small, a few are medium and a few are large. The balls in A and B are put into a box, C. What is the number of balls in C which are neither large nor small?*

- *(g) A box contains about 20 balls of various sizes. There are many more large balls than small balls. What is the number of small balls?*

- *Traditionally, such problems are dismissed as ill-posed, and are viewed as off limits to mathematics. What is not recognized is that this need not be the case. Through employment of concepts and techniques drawn from CWW, an important capability can be added to mathematics—the capability to construct mathematical solutions of computational problems which are stated in a natural language.*

12/263

- *Traditional mathematics does not have this capability. The importance of this capability derives from the fact that much of human knowledge and, in particular, world knowledge, is described in a natural language. The capability will grow in importance and visibility as we move further into the age of automation of everyday reasoning and decision-making.*

13/263

NOTE

• *When A poses a CNL problem to B, B is likely to be unsure of how imprecise terms such as most, tall, usually, etc. should be interpreted. If this is the case, B can ask A to precisiate the meaning of such terms. In constructing mathematical solutions to CNL problems, what is assumed is that the imprecise terms are labels of fuzzy sets with specified membership functions.*

• *If the membership functions are not specified, they can be elicited from A. As an illustration, assume that B wants to elicit from A the membership function of young. To this end, A can ask B a sequence of questions of the form: To what degree does age u fit your perception of young? u is a numerical value of age, say 25, which ranges over possible values of age.*

- *This mode of explicitation exploits the remarkable human capability to graduate perceptions.*

BASIC STRUCTURE OF CWW

- *The point of departure in CWW is a question, q, of the form: What is the value of a variable, Y? The answer to this question is expected to be derived from a collection of propositions, I, $I=(p_1, ..., p_n)$, which is referred to as the information set. Generally, at least some of the propositions in I are drawn from a natural language.*

- *I is closed if addition to I of propositions drawn from external sources of information, in particular, from world knowledge, is not allowed. Otherwise, I is open. An added proposition is denoted as +p.*
- *In essence, I is a collection of question-relevant propositions.*

18/263

- *The terminus consists of an answer of the form: Y is ans(q/I). Generally, ans(q/I) is not a value of Y but a restriction (generalized constraint) on the values which Y is allowed to take (Zadeh 2006a). Equivalently, ans(q/I) identifies those values of Y which are consistent with I. In CWW, consistency is equated to possibility, with the understanding that possibility is a matter of degree.*

19/263

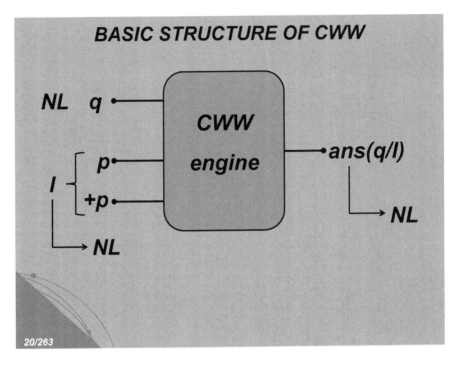

BASIC STRUCTURE OF CWW

20/263

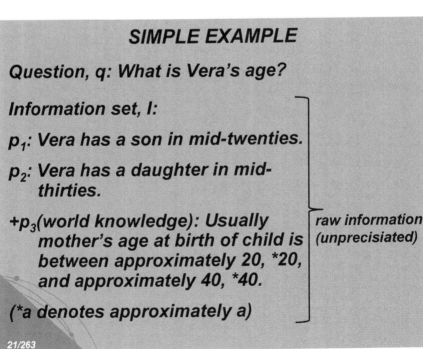

SIMPLE EXAMPLE

Question, q: What is Vera's age?

Information set, I:

p_1: *Vera has a son in mid-twenties.*

p_2: *Vera has a daughter in mid-thirties.*

$+p_3$*(world knowledge): Usually mother's age at birth of child is between approximately 20, *20, and approximately 40, *40.*

(*a denotes approximately a)

raw information (unprecisiated)

21/263

EXAMPLE—VERA'S AGE

- First step, partial precisiation of the information set, I*:

p_1*: Age(Son) is *25

p_2*: Age(Daughter) is *35

+p_3*: Age(Mother.at.birth) is usually [*20,*40]

- ans(q/I): Age(Vera) is

(*25+usually[*20,*40]) $\wedge$ (conjunction) (*35+usually[*20,*40])

- Note. For simplicity "usually" is interpreted as "always."

22/263

KEY IDEA

- A prerequisite to computation is precisiation of meaning. Raw (unprecisiated) natural language cannot be computed with. Precisiation of meaning has a position of centrality in CWW.

- Understanding of meaning is a prerequisite to precisiation of meaning.

23/263

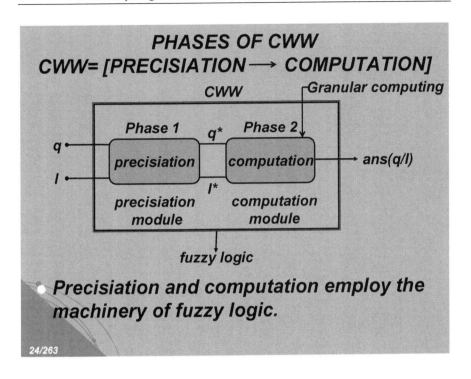

PHASES OF CWW

CWW= [PRECISIATION $\longrightarrow$ COMPUTATION]

- **Precisiation and computation employ the machinery of fuzzy logic.**

24/263

NOTE

- **To facilitate understanding of the basic ideas which underlie CWW, a clarification dialogue is included in the Appendix.**

25/263

LEVELS OF COMPLEXITY IN CWW

- *There are two principal levels of complexity in CWW: Level 1 CWW (basic CWW) and Level 2 CWW (advanced CWW). Complexity is related to the difficulty of construction of computational models of propositions in I.*

26/263

LEVELS OF COMPLEXITY IN CWW

- *Informally, in Level 1 CWW (CWW1) the objects of computation/precisiation are words, phrases and simple propositions.*
- *Informally, in Level 2 CWW (CWW2) the objects of computation/precisiation are possibly complex propositions.*

27/263

EXAMPLES

- **Level 1 (basic):**

  Robert is young

  X is small

  If X is small then Y is large

- **Level 2 (advanced):**

  It is very unlikely that there will be a significant decrease in the price of oil in the near future.

NOTE

Historically, Level 2 CWW may be viewed as a sequel to and a generalization of Level 1 CWW. Today, it is Level 1 CWW that is in preponderant use and a center of research activity. In coming years, the intrinsic importance of Level 2 CWW is likely to gain recognition, leading to a wide variety of new application areas.

NOTE

- *Generally, Level 2 CWW is needed for construction of mathematical solutions of computational problems which are stated in a natural language.*

- *Linguistic variables and fuzzy if-then rules (Takagi and Sugeno 1985, Bardossy and Duckstein 1995, Yen and Langari 1999, Mendel 2001) fall within the province of Level 1 CWW.*

NOTE

- *In part, linguistic variables and fuzzy if-then rules are widely used because the underlying theory is easy to understand and easy to apply.*

NOTE

- *Level 1 CWW is associated with an extensive literature and a wide range of applications. Level 2 CWW is not, at least at this stage of its development. For this reason, exposition of CWW in this presentation is focused on Level 2 CWW. More specifically, what will not be discussed is the widely used machinery of linguistic variables and fuzzy if-then rules.*

32/263

IMPRECISION OF NATURAL LANGUAGES

- *A natural language is basically a system for describing perceptions.*
- *Perceptions are intrinsically imprecise, reflecting the bounded ability of human sensory organs and ultimately the brain, to resolve detail and store information. Imprecision of perceptions is passed on to natural languages.*

33/263

IMPRECISION OF NATURAL LANGUAGES AND FUZZY LOGIC

- *The principal source of imprecision in natural languages is unsharpness of boundaries of classes which underlie the meaning of words.*

- *Unsharpness of class boundaries = fuzziness.*

- *Fuzzy logic is the logic of classes with unsharp boundaries.*

34/263

- *Fuzzy set= precisiated (graduated) class with unsharp boundaries.*

- *Graduation (precisiation)= association of a class, A, which has unsharp boundaries, with a scale of degrees— more concretely, with a membership function. Degrees are allowed to be fuzzy (fuzzy sets of type 2). (Zadeh 1975a, Mendel 2001) In Level 2 CWW, fuzziness of degrees is the rule rather than exception.*

35/263

IMPORTANT POINTS

- *Most words (phrases) in a natural language are labels of classes with unsharp boundaries. Examples: near, soft, fast, tall, hand, etc. Fuzziness of words is a concomitant of fuzziness of perceptions.*

- *Words(phrases) are precisiated through graduation.*

EXAMPLE—GRADUATION OF MIDDLE-AGE

- *Imprecision of meaning = fuzziness of meaning*
- *Computational model of middle-age (trapezoidal fuzzy set)*

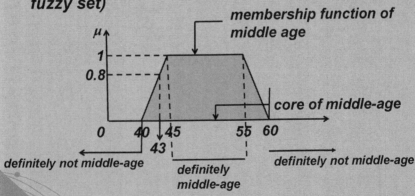

- *Note: Parameters are context-dependent*

EXAMPLE—HONDA FUZZY LOGIC TRANSMISSION

Fuzzy Sets

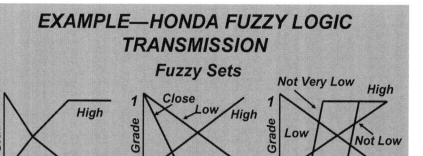

Control Rules:

1. If (speed is low) and (shift is high) then (-3)
2. If (speed is high) and (shift is low) then (+3)
3. If (throt is low) and (speed is high) then (+3)
4. If (throt is low) and (speed is low) then (+1)

38/263

PRINCIPAL RATIONALES FOR COMPUTING WITH WORDS

Basic premises:

- **Words are less precise than numbers**
- **Precision carries a cost**
- **Numbers are respected, words are not**

- **Rationale A.**

  Use words when numbers are not known or are too costly to obtain. Use of words is a necessity.

39/263

● *Rationale B.*

Words are good enough.

Numbers are known but there is a tolerance for imprecision which can be exploited by employing words in place of numbers, aiming at a reduction in cost and achieving simplicity. Use of words is advantageous.

40/263

● *Rationale C.*

Linguistic summarization.

Words are used to summarize numerical information. Use of words is expedient.

41/263

RATIONALES FOR THE USE OF WORDS

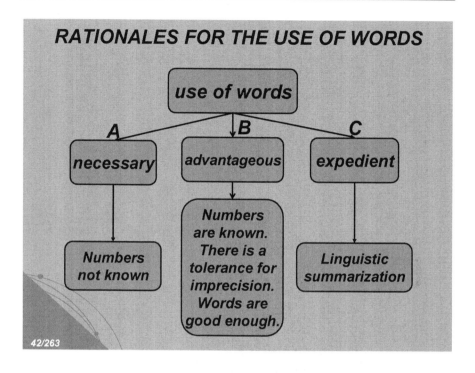

42/263

NOTE

- Today, most applications of CWW1, especially in the realm of consumer products, are based on Rationales B and C. A key role is played by linguistic summarization.

43/263

LINGUISTIC SUMMARIZATION VIA GRANULATION (FUZZY PARTITION)

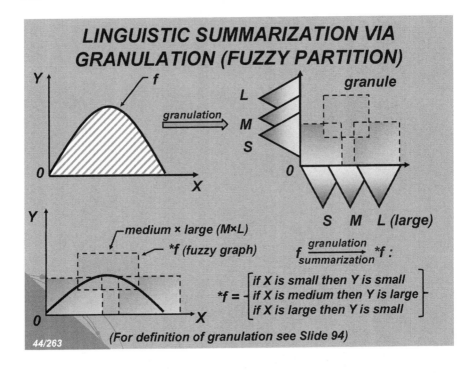

(For definition of granulation see Slide 94)

44/263

FUZZY LOGIC GAMBIT

- **Fuzzy Logic Gambit = deliberate imprecisiation through granulation, followed by graduation**

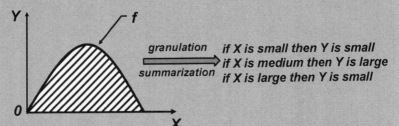

- **The Fuzzy Logic Gambit is employed in many applications of fuzzy logic, especially in the realm of consumer products**

45/263

NATURAL-LANGUAGE-BASED (LINGUISTIC) SYSTEM MODELIZATION (LEVEL1 CWW)

I (information set)

system ⟶ M(S) ⟶ CWW

linguistic model

- *Example: Yamakawa's inverted pendulum (Yamakawa 1989), employing a linguistic model of a system.*

46/263

YAMAKAWA'S INVERTED PENDULUM (1989)

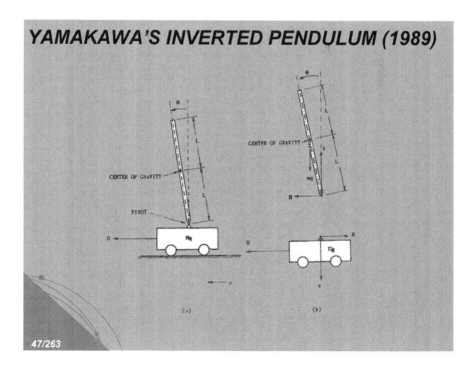

47/263

MATHEMATICAL MODEL (v-PRECISE)

$$I\ddot{\theta} = VL\sin\theta - HL\cos\theta,$$

$$V - mg = -mL(\ddot{\theta}\sin\theta + \dot{\theta}^2\cos\theta),$$

$$H = m\ddot{y} + mL(\ddot{\theta}\cos\theta - \dot{\theta}^2\sin\theta),$$

$$U - H = M\ddot{y},$$

48/263

PL : **P**ositively **L**arge   NL : **N**egatively **L**arge
PM : **P**ositively **M**edium   NM : **N**egatively **M**edium
PS : **P**ositively **S**mall   NS : **N**egatively **S**mall
              ZR : Approximately **Z**e**r**o

Assignment of Membership Functions

49/263

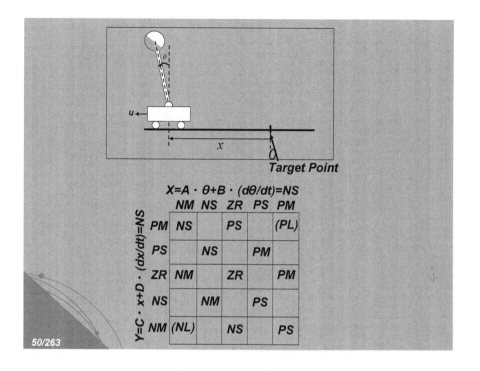

Target Point

$$X = A \cdot \theta + B \cdot (d\theta/dt) = NS$$

$Y=C \cdot x+D \cdot (dx/dt)=NS$		NM	NS	ZR	PS	PM
	PM	NS		PS		(PL)
	PS		NS		PM	
	ZR	NM		ZR		PM
	NS		NM		PS	
	NM	(NL)		NS		PS

50/263

LINGUISTIC MODEL (PERCEPTION-BASED)

IF θ is PM AND θ̇ is ZR, THEN ẏ is PM,
ALSO

IF θ is PS AND θ̇ is PS, THEN ẏ is PS,
ALSO

IF θ is PS AND θ̇ is NS, THEN ẏ is ZR,
ALSO

IF θ is NM AND θ̇ is ZR, THEN ẏ is NM,
ALSO

IF θ is NS AND θ̇ is NS, THEN ẏ is NS,
ALSO

IF θ is NS AND θ̇ is PS, THEN ẏ is ZR,
ALSO

IF θ is ZR AND θ̇ is ZR, THEN ẏ is ZR.

51/263

HISTORICAL NOTE

- *In science, there is a deep-seated tradition of according much more respect to numbers than to words. The countertraditional spirit of the advocacy of the use of words evoked criticism and derision rather than approbation. What was not recognized is that the deliberate sacrifice of precision is a gambit—a gambit which opens the door to important applications.*

- *A typical negative view was that of Rudolf Kalman, a brilliant system scientist and a good friend.*

52/263

COMMENT

(Kalman 1972)

I would like to comment briefly on Professor Zadeh's presentation. His proposals could be severely, ferociously, even brutally criticized from a technical point of view. This would be out of place here. But a blunt question remains:

Is Professor Zadeh presenting important ideas or is he indulging in wishful thinking?

53/263

The most serious objection to fuzzification of system analysis is that lack of methods of system analysis is not the principal scientific problem in the systems field. That problem is one of developing basic concepts and deep insight into the nature of systems, perhaps trying to find something akin to the laws of Newton. In my opinion, Professor Zadeh's suggestions have no chance to contribute the solution of this basic problem.

54/263

To take a concrete example, modern experimental research has shown that the brain, far from fuzzy, has in many areas a highly specific structure. Progress in brain research is now most rapid in anatomy where the electron microscope is the new tool clarifying regularities of structure which previously were seen only in a fuzzy way.

No doubt Professor Zadeh's enthusiasm for fuzziness has been reinforced by the prevailing political climate in the U.S.- one of unprecedented permissiveness.

Fuzzification is a kind of scientific

55/263

permissiveness; it tends to result in socially appealing slogans unaccompanied by the discipline of hard scientific work and patient observation. I must confess that I cannot conceive of fuzzification as a viable alternative for the scientific method; I even believe that it is healthier to adhere to Hilbert's naïve optimism, Wir wollen wissen: wir werden wissen.

It is very unfair for Professor Zadeh to present trivial examples where fuzziness is tolerable or even comfortable and in any case irrelevant, and then imply, though

not formally claim, that his vaguely outlined methodology can have an impact on deep scientific problems. In any case, if the fuzzification approach is going to solve any difficult problems, this is yet to be seen.

Zadeh, in response:

To view Professor Kalman's rather emotional reaction to my presentation in a proper perspective, I should like to observe that, up to a certain point in time, Professor Kalman and I have been traveling along the same road, by which I mean that both of us believed in the power of mathematics, in the eventual triumph of logic and precision over vagueness. But then I made a right turn, or maybe even started turning backwards, whereas Professor Kalman has stayed on the same road. Thus, today, I no loner believe, as Professor Kalman does, that the solution to the kind of problems referred to in my talk lies within the conceptual framework of classical mathematics. In taking this position, I realize, of course, that I am challenging scientific dogma—the dogma that was alluded to by Professor Caianiello in his remarks.

Now, when one attacks dogma, one must be prepared to become the object of counterattack on the part of those who believe in the status quo. Thus, I am not surprised when in reaction to my views I encounter not only enthusiasm and approbation, but also criticism and derision. Nevertheless, I believe that, in time, the concepts that I have presented will be accepted and employed in a wide variety of areas. Their acceptance, however, may be a grudging one by those who have been conditioned to believe that human affairs can be analyzed in precise mathematical terms. The pill, though a bitter one, will be swallowed eventually. Indeed, in retrospect, the unconventional ideas suggested by me may well be viewed as self-evident to the point of triviality. Thank you.

KEY POINT

- *What Kalman and other critics of CWW did not realize is that precisiation of meaning opens the door to construction of a system of computation which offers an important capability that traditional systems of computation do not have—the capability to compute and reason with information described in a natural language.*

A RELATED KEY POINT

● *When we discuss precisiation of meaning it is important to differentiate between precision in value and precision in meaning. Traditionally, precision relates to precision of value. In fuzzy logic, precision relates, in the main, to precision in meaning. As a simple illustration, assume that the value of X, a real-valued variable, is not known precisely and is described as*

62/263

small, where small is a fuzzy set with a specified membership function. This description is value-wise imprecise but is meaning-wise precise. More generally, what can be said is that, fundamentally, fuzzy logic moves from precision in value to precision in meaning. This is a key idea that underlies fuzzy logic and defines its spirit. In this perspective, fuzzy logic may be viewed as a precise logic of imprecision.

63/263

NOTE

- *In addition to playing a pivotal role in CWW, precisiation of meaning is important in its own right. More concretely, precisiation of meaning is needed in a variety of fields, among them legal reasoning, wording of contracts and specifications, definition of concepts, human-machine communication, etc. For such applications, CWW offers an effective method of precisiation.*

NOTE

- *It is of interest to note that the concept of precisiation, in the sense in which it is used in CWW, does not exist within linguistics or computational linguistics.*

SUMMARY

- *CWW= precisiation of semantic entities, mainly words, phrases and propositions, followed by computation with precisiated semantic entities (computational/mathematical models of semantic entities.)*

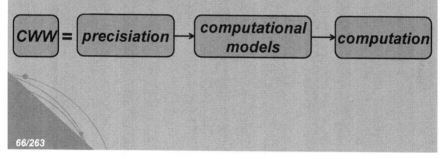

66/263

VISUAL REPRESENTATION OF RESTRICTIONS— THE CONCEPT OF A Z-MOUSE

- *A Z-mouse is an electronic implementation of a spray pen. The cursor is a round fuzzy mark called an f-mark. The color of the mark is a matter of choice. A dot identifies the centroid of the mark. The cross-section of an f-mark is a trapezoidal fuzzy set with adjustable parameters.*

Example. Vera's Age

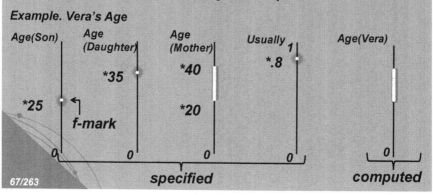

67/263

I believe that Robert is very honest

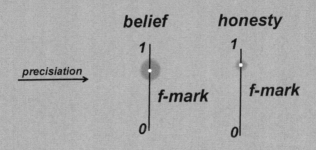

belief   honesty

precisiation

f-mark   f-mark

68/263

MORE ON Z-MOUSE

- If I am not sure what the degree is, and I am allowed to use a Z-mouse, I will put a fuzzy f-mark on the scale.

- A fuzzy f-mark reflects imprecision of my perception.

- A Z-mouse reads my f-mark and represents it internally into a trapezoidal fuzzy set—a fuzzy set which serves as an object of computation for the machinery of Computing with Words.

69/263

EXAMPLE OF APPLICATION OF Z-MOUSE

- *I am scheduled to fly from San Francisco to Los Angeles. My flight is scheduled to leave at 5pm. I have to be at the airport about an hour before departure. Usually it takes about forty five minutes to get to the airport from my home. I would like to be pretty sure that I arrive at the airport in time. At what time should I leave my home?*

70/263

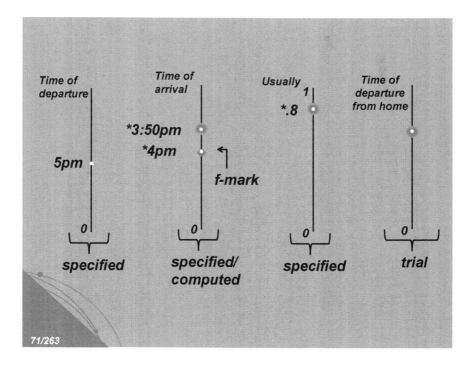

71/263

COMPUTING WITH WITH WORDS— HOW?

PHASE 1— PRECISIATION

Restriction-based Semantics (RS)

PRECISIATION OF MEANING

- Basically, precisiation involves construction of computational/ mathematical models of words, phrases, propositions, questions and other types of semantic entities.
- Precisiation of meaning goes beyond representation of meaning.

L.A. Zadeh: Computing with Words, STUDFUZZ 277, p. 41–72.
springerlink.com

SIMPLE EXAMPLES OF PRECISIATION

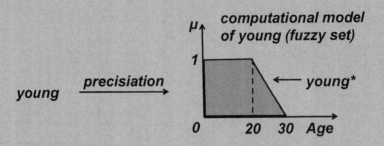

Note: Parameters of the trapezoidal membership function are context-dependent

- **Most Swedes are tall**

precisiation *Proportion(tall. Swedes/Swedes) is most*

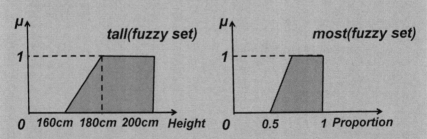

- **Precisiation of words is followed by composition.**

BASIC STRUCTURE OF PRECISIATION

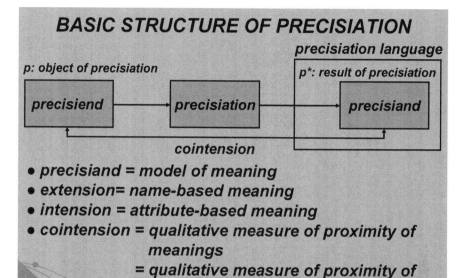

precisiation language

p: object of precisiation

p*: result of precisiation

precisiend → precisiation → precisiand

cointension

- precisiand = model of meaning
- extension= name-based meaning
- intension = attribute-based meaning
- cointension = qualitative measure of proximity of meanings
  = qualitative measure of proximity of the model (precisiand) and the object of modeling (precisiend)

77/263

PRECISIATION = MODELIZATION

- Let p be a proposition drawn from a natural language. In CWW, precisiation of p involves construction of a computational/mathematical model of p. In this perspective, precisiation may be viewed as a form of modelization.

- A desideratum of precisiation/modelization is high cointension.

78/263

COINTENSION PRINCIPLE

- *A precisiend has a multiplicity of precisiands.*
- *Generally, achievement of cointensive precisiation requires that if the precisiend is fuzzy so must be the precisiand.*
- *Crisp definitions of fuzzy concepts is the norm in science. What is widely unrecognized is that, in general, crisp definitions of fuzzy concepts are not cointensive.*

79/263

THE CONCEPT OF A PROPOSITION

- *The concept of a proposition is one of the most basic concepts in the realms of both natural and synthetic languages.*
  A dictionary definition of a proposition reads: An expression in language or signs of something that can be believed, doubted, or denied or is either true or false.
- *Familiar examples are: Robert is very bright, Leslie is much taller than Ixel, most Swedes are tall, etc.*

80/263

HOW IS THE CONCEPT OF A PROPOSITION VIEWED IN CWW?—THE MEANING POSTULATE

- In CWW, conventional definitions of propositions are abandoned.
- Instead, a proposition, p, is defined in a way that lends itself to construction of a computational model of p.
- Premises

(a) Information=restriction

(b) A proposition, p, is a carrier of information.

THE MEANING POSTULATE (MP)

- The meaning postulate is a concomitant of premises (a) and (b). Let X be a variable which is associated with (implicit in p). The meaning postulate, MP, may be expressed as:

$$MP: p = \text{restriction on } X$$

- The meaning postulate is the point of departure in precisiation of meaning in CWW.

THE CONCEPT OF A RESTRICTION— A CLOSER LOOK

- *I am asked: What is the value of a real-valued variable X? My answer is: I do not know the value precisely but I have a perception which I can express as a restriction (generalized constraint) (Zadeh 2006a) on the values which X can take.*

EXAMPLES

- $8 \leq X \leq 10$
- *X is small*
- *Usually X is small*
- *X is normally distributed with mean 9 and variance 2.*
- *It is likely that X is between 8 and 10.*

REPRESENTATION OF A RESTRICTION

● *A restriction (generalized constraint), R(X), may be represented as:*

$$R(X): \quad X \text{ isr } R$$

where X is the restricted (constrained) variable, R is the restricting (constraining) relation and r is an indexical variable which defines how R restricts X.

RESTRICTIONS

● *There are many different kinds of restrictions. The principal restrictions are: possibilistic (r=blank); probabilistic (r=p) and combinations of possibilistic and probabilistic restrictions. A special case of a combination of possibilistic and probabilistic restrictions is a Z-restriction (r=z) (Zadeh 2011)*

EXAMPLES

- *Possibilistic restriction (r=blank):*

$$R(X): \quad X \text{ is } A$$

where A is a fuzzy set in U with the membership function μ_A. A plays the role of the possibility distribution of X (Zadeh 1978)

$$Poss(X=u)= \mu_A(u)$$

PROBABILISTIC RESTRICTION

- *Probabilistic restriction (r=p):*

$$R(X): \quad X \text{ isp } P$$

where P plays the role of the probability distribution of X.

$$Prob(u \leq X \leq u+du)=p(u)du$$

where p is the probability density function of X.

Z-RESTRICTION

- *Z-restriction (r=z):*

$$R(X): \quad X \ isz \ Z$$

where Z is a combination of possibilistic and probabilistic restrictions defined by:

$$Prob(X \ is \ A) \ is \ B$$

in which A and B are fuzzy sets.

89/263

- *Usually, A and B are labels drawn from a natural language. A Z-restriction may be expressed as:*

$$X \ isz \ (A,B)$$

(A,B) is referred to as a Z-number (Zadeh 2011)

90/263

EXAMPLES OF Z-RESTRICTIONS

- Usually temperature is low $\longrightarrow$ temperature isz (low, usually)

- Probably John is tall $\longrightarrow$ Height(John) isz (tall, probable).

- Important note

Usually X is A,

where A is a fuzzy set, is a Z-restriction

91/263

DIRECT AND INDIRECT RESTRICTIONS

- A restriction is direct if it is of the form:

$$R(X):\quad X\ isr\ R$$

- A restriction is indirect if it is of the form:

$$R(X):\quad f(X)\ isr\ R$$

where f is a specified function or functional.

92/263

EXAMPLE OF INDIRECT RESTRICTION

$R(X):\int_R \mu(u)p(u)du$ is likely

is an indirect restriction on p.

- Note: The term "restriction" is sometimes applied to R.

93/263

COLLECTION OF RESTRICTIONS—GRANULATION

- A variable, X, may be associated with a collection of restrictions. Generally, this is the case when X is granulated, that is, X is a granular variable. A linguistic variable is a granular variable in which the labels of granules are drawn from a natural language. A simple example of a linguistic variable is the following.

94/263

GRANULATION OF AGE (LINGUISTIC VARIABLE)

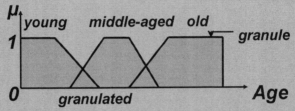

Granulation plays a pivotal role in fuzzy logic (Zadeh 1997) and, especially, in granular computing. (Zadeh 1998, Lin 1998, Pedrycz and Bargiela 2002) Linguistic variables are employed extensively in applications of fuzzy logic.

NOTE

- Note. A granule is a restriction but not every restriction is a granule. Informally, a granule is a restriction in which the values which X can take are drawn together by similarity, proximity or functionality.

PRINCIPAL LEVELS OF GENERALITY OF RESTRICTIONS

Z-numbers ———————— *Level 3*

fuzzy numbers   *random numbers* — *Level 2*

intervals ———————— *Level 1*

real numbers ———————— *ground level*

97/263

RESTRICTION-BASED SEMANTICS

- In CWW, precisiation of propositions is carried out through the use of what is referred to as restriction-based semantics, RS.
- The point of departure—and a key idea in RS—is that of representing the precisiated meaning of a proposition, p, as a restriction

$$p \xrightarrow{precisiation} X \text{ isr } R$$

- The expression X isr R is referred to as a canonical form of p.

98/263

EXAMPLES OF CANONICAL FORMS

- p: Robert is young⟶ Age(Robert) is young
  $$\underset{X}{\uparrow}\quad \underset{blank}{\uparrow}\quad \underset{R}{\uparrow}$$

- p: Most Swedes are tall ⟶
  Proportion(tall Swedes/Swedes) is most
  $$\underset{X}{\uparrow}\qquad\qquad\qquad \underset{R}{\uparrow}$$

- p: Usually it takes Robert about an hour to get home from work ⟶
  Travel.time.from.office.to.home (Robert) isz
  (approximately 1 hr, usually)   X
  $$\underset{R\ (Z\text{-}number)}{\uparrow}$$

99/263

NOTE

- Equating a proposition drawn from a natural language to a restriction bridges the divide between linguistics and mathematics.

100/263

NOTE

- *X need not be a scalar variable.*

Example:

- *p: Robert gave a ring to Anne, X may be represented as the 3-tuple (Giver, Recipient, Object), with the corresponding values of R being (Robert, Anne, Ring).*

101/263

NOTE

- *A semantic network may be viewed as a canonical form of p, with X as an n-ary variable.*

102/263

KEY IDEA

Proposition = Restriction

- Given p, three basic questions arise:
1. What is the variable, X, which is restricted?
2. What is the restricting relation, R?
3. How does R restrict X?

103/263

KEY POINTS

- The answers to these questions define the meaning of p.
- Typically, in the case of propositions drawn from a natural language, X and R are implicit (hidden).
- Generally, X and R are identified by inspection. Choice of X is influenced by world knowledge.

104/263

KEY POINTS

- *R is determined by X.*
- *X and R are referred to, respectively, as the focal variable and the focal relation.*
- *X and R are analogous to the subject and predicate of a sentence.*

NOTE

- *In general, the focal variable, X, is not unique. However, it is usually the case that among possible choices either there is one that has higher plausibility than others, or there are a few that are closely related. For example, if*

  p: Leslie is much taller than Ixel

  then a plausible choice of X is

  X: Height(Leslie)

in which case the corresponding constraining relation is

    R: Much taller than Ixel.

Another plausible choice is

    X: (Height(Leslie), Height(Ixel)).

Correspondingly,

    R: Much taller

107/263

NOTES

- *Restriction-based semantics is rooted in test-score semantics (Zadeh 1981)*
- *Restriction-based semantics may be viewed as a generalization of traditional approaches to semantics of natural languages—mainly possible-world semantics and truth-conditional semantics (Lambert 1970)*
- *Restriction-based semantics has a far greater precisiation capability than possible-world and truth-conditional semantics.*

108/263

- *RS is not needed for Level 1 (basic) CWW. In particular, what is important to note is that RS is not needed for computation with linguistic variables and fuzzy if-then rules.*
- *The basics of RS are discussed in the following.*

CANONICAL FORM of p: CF(p)

- *When the meaning of p is represented as a restriction, the restriction is the canonical form of p, CF(p). Thus,*

$$CF(p): X \text{ isr } R$$

- *The concept of a canonical form of p has a position of centrality in precisiation of meaning of p.*

NOTE

- *It is important to note that representing a proposition as a restriction is greatly facilitated by the fact that restrictions in a natural language are predominantly possibilistic. Possibilistic restrictions are easiest to compute with.*

THE CONCEPT OF AN EXPLANATORY DATABASE

- *Constructing the canonical form of p is merely a first step in precisiation of p, since X and R are expressed in a natural language and hence require precisiation. What is needed for precisiation of X and R is the concept of an explanatory database. The concept of an explanatory database and its application to precisiation of X and R are described in the following.*

THE CONCEPT OF EXPLANATORY DATABASE (ED)

- *In restriction-based semantics, the concept of an explanatory database, ED, serves as a basis for precisiation of meaning of p (Zadeh 1983a). More concretely, ED is a collection of relations, with the names of relations drawn, but not exclusively, from the constituents of p. Instantiated ED is denoted as ED^+.*

113/263

- *Basically, ED may be viewed as the information which is needed to define X and R. Alternatively, ED may be viewed as the information which is needed to assess the truth-value of p. There is a connection between the concept of ED and the concept of a model in model theory (Chang and Keisler 1973)*

  Example. For the proposition, p: Most Swedes are tall, ED may be represented as:

114/263

NOTE

ED=POPULATION.SWEDES[Name; Height]+TALL[Height;μ]+ MOST[Proportion;μ],

where + plays the role of a comma.

- *It is important to note that precisiation of X and R is needed because X and R are described in a natural language.*

ADDITIONAL EXAMPLE OF ED

- *Note. This and other examples are discussed in greater detail in Zadeh 1983a.*
- *p: Brian is much taller than most of his friends.*

X: Height of Brian.

R: Much taller than most of his friends.

ED = HEIGHT[Name; Height]+ FRIENDS.BRIAN[Name; μ]+

MUCH.TALLER [Height1; Height2; μ]+
MOST[Proportion; μ]

In FRIENDS.BRIAN, μ is the degree to which Name is a friend of Brian.

- In relation to possible-world semantics, ED⁺ may be viewed as the description of a possible world.

117/263

THE CONCEPT OF A PRECISIATED CANONICAL FORM, CF*(p)

- After X and R have been identified and the explanatory database, ED, has been constructed, X and R may be defined as functions of ED. As was noted earlier, definitions of X and R may be viewed as precisiations of X and R. Precisiated X and R are denoted as X* and R*, respectively.

118/263

- A canonical form, CF*(p), with precisiated values of X and R, X* and R*, will be referred to as a precisiated canonical form, CF*(p): X* isr R*.

- In RS, the precisiated canonical form of p is equated to the meaning of p.

- In the following, construction of the precisiated canonical form of p is discussed in greater detail.

FROM p TO CF*(p): X* isr R*

MEANING = PRECISIATED CANONICAL FORM

- *The concepts discussed so far provide a basis for a relatively straightforward procedure for constructing the precisiated canonical form of a given proposition, p. As was noted earlier, the precisiated canonical form may be viewed as a computational/mathematical model of p. Effectively, the precisiated canonical form may be interpreted as a representation of precisiated meaning of p.*

121/263

- *A summary of the procedure for computing the precisiated canonical form of p is presented in the following.*

122/263

CLARIFICATION OF MEANING

- *A preliminary step is that of clarification, if needed, of the meaning of the given proposition. This step requires world knowledge.*

Examples:

- *Overeating causes obesity* $\xrightarrow{clarification}$
  Most of those who overeat are obese.
- *Obesity is caused by overeating* $\xrightarrow{clarification}$
  Most of those who are obese, overeat.

123/263

- *Young men like young women* $\xrightarrow{clarification}$
  Most young men like mostly young women.

- *Swedes are much taller than Italians*
  $\xrightarrow{clarification1}$ *Most Swedes are much taller than most Italians.*
  $\xrightarrow{clarification2}$ *The average height of Swedes is much greater than the average height of Italians.*

124/263

- *Clarification (disambiguation) of a predicate.*

Most tall Swedes

$\xrightarrow{\text{clarification1}}$ *Mostly tall Swedes*

$\xrightarrow{\text{clarification2}}$ *Most of tall Swedes*

SUMMARY OF PRECISIATION PROCEDURE

- *Step 1. Identification (explicitation) of X and R.*

Identify the restricted (focal) variable, X, and the corresponding restricting (focal) relation, R. R depends on X. Generally, X and R are identified by inspection.

- **Step 2. Construction of ED.**
  What information is needed to precisiate (define) X and R? An answer to this question identifies the explanatory database, ED. Equivalently—as was noted earlier—ED may be viewed as an answer to the question: What information is needed to compute the truth-value of p?

- **Step 3. Precisiation of X and R.**
  How can the information in ED be used to precisiate the values of X and R? This step leads to precisiated values of X and R, X* and R*, and thus results in the precisiated canonical form, CF*(p). X* and R* are constructed by inspection.

- *Precisiated X* and R* are expressed as functions of ED.*

KEY CONCEPTS

- More concretely, X^* and R^* may be expressed as:

$$X^* = f(ED)$$

$$R^* = g(ED)$$

- The precisiated canonical form, $CF^*(p)$: X^* isr R^*, induces a restriction on ED which will be referred to as the generalized intension of p, $\mu(p)$.

- The generalized intension of p may be viewed as a representation of a deep semantic structure of p.

129/263

- $\mu(p)$ may be interpreted as the possibility distribution of ED given p.

- Equivalently, $\mu(p)$ may be interpreted as the truth-value of p given ED^+.

- $\mu(p)$ plays the role of a computational/mathematical model of p. (Zadeh 1981)

- End of precisiation procedure.

130/263

SUMMARY OF PROCEDURE

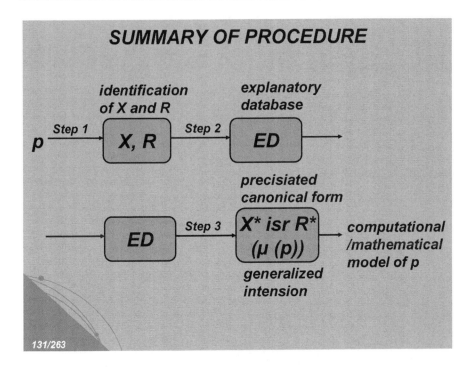

131/263

NOTE
THE CONCEPT OF A PROTOFORM

- *A protoform, (prototypical form) of p, denoted as Pr(p), is an abstracted version of the generalized intension of p (Zadeh 2006a). Pr(p) represents an abstracted deep semantic structure of p. Simple examples.*

  Pr(Robert is tall)=A(B) is C

  Pr(Most Swedes are tall)=QA's are B's

132/263

- *The concept of a protoform plays an important role in computation with precisiated propositions, and in search and question-answering systems (Zadeh 2006b).*

- *It is important to note that in fuzzy logic the rules of deduction involve protoforms.*

133/263

CONCLUSION

- *In RS, a precisiated meaning of a proposition, p, is equated to its precisiated canonical form, $CF^*(p)$: X^* isr R^*, with the understanding that X^* isr R^* is described as a procedure.*

- *The precisiated canonical form translates into a restriction on the explanatory database, ED. The generalized intension, $\mu(p)$, plays the role of a computational/mathematical model of p.*

134/263

- *Representation of p as a restriction is the centerpiece of restriction-based semantics, RS. Through precisiation of meaning, RS opens the door to a wide-ranging enlargement of the role of natural languages in scientific theories and engineering systems. Importantly, RS plays a pivotal role in empowering mathematics to construct mathematical solutions to computational problems stated in natural languages.*

PHASE 2— COMPUTATION

COMPUTATION WITH RESTRICTIONS

- *In CWW, through representation of the meaning of a proposition as a restriction, the problem of computation with information described in natural language reduces to the problem of computation with restrictions. As was noted earlier, in the realm of natural languages restrictions are for the most part possibilistic.*

L.A. Zadeh: Computing with Words, STUDFUZZ 277, p. 73–89.
springerlink.com

For this reason, in the following attention is focused on computation with possibilistic restrictions.

138/263

FROM PRECISIATION TO COMPUTATION

Phase 1

$$I \begin{cases} p_1 \\ \vdots \\ p_{n-1} \\ P_n \\ q \end{cases} \rightarrow \boxed{precisiation} \rightarrow \begin{cases} p_1^*: X_1 \; isr_1 \; R_1 \\ \vdots \\ p_{n-1}^*: X_{n-1} \; isr_{n-1} \; R_{n-1} \\ p_n^*: X_n \; isr_n \; R_n \\ q^* \end{cases} I^*$$

Phase 2

$$I^* \begin{cases} X_1 \; isr_1 \; R_1 \\ \vdots \\ X_{n-1} \; isr_{n-1} \; R_{n-1} \\ X_n \; isr_n \; R_n \\ q^* \end{cases} \rightarrow \boxed{\begin{array}{c} computation \\ with \; restrictions \end{array}} \rightarrow ans(q/I)$$

139/263

• *In large measure, computation with restrictions (generalized constraints) involves the use of rules which govern propagation and counterpropagation of restrictions (Calculus of fuzzy restrictions, Zadeh 1974). Among such rules, the principal rule is the extension principle (Zadeh 1965, 1975a, 2011).*

• *There are many versions of the extension principle. Let Y be a function of $X_1, ..., X_n$, $Y=f(X_1, ..., X_n)$. Basically, an extension principle is a rule which governs computation of the restriction on Y, R(Y), given restrictions on $X_1, ..., X_n$.*

NOTE

- *The calculi of fuzzy if-then rules in Level 1 CWW may be viewed as special cases of propagation of possibilistic restrictions.*

EXTENSION PRINCIPLE (POSSIBILISTIC)

- *X is a variable which takes values in U, and f is a function from U to V. The point of departure is a possibilistic restriction on f(X) expressed as f(X) is A, where A is a fuzzy set in V which is defined by its membership function $\mu_A(v)$, $v \varepsilon V$.*

- *g is a function from U to W. The possibilistic restriction on f(X) induces a possibilistic restriction on g(X) which may be expressed as g(X) is ?B, where B is a fuzzy set in W. The question is: What is B? In symbols,*

$$\frac{f(X) \text{ is } A}{g(X) \text{ is } ?B}$$

The answer to this question is the solution of a variational problem expressed as:

$$\mu_B(w) = sup_u\, \mu_A(f(u))$$

subject to

$$w = g(u)$$

where μ_A and μ_B are the membership functions of A and B, respectively.

144/263

Equivalently, the possibilistic extension principle may be expressed as:

$$\frac{\begin{array}{l}?Y{=}g(X)\\ f(X) \text{ is } A\end{array}}{\mu_Y(v) = sup_u\, \mu_A(f(u))}$$

subject to

$$v = g(u)$$

145/263

NOTE

- *In a more general setting, the extension principle is concerned with propagation of restrictions—restrictions on both functions and their arguments. Schematically,*

$$Z = g(X, Y)$$

- *There are many different ways in which X, Y and g may be restricted. Only a few have been explored.*

A SUMMARY OF COMPUTATION OF ans(q/I)

- *For convenience, in the following the information set, I, is represented as a composite proposition, $p=(p_1, ..., p_n)$.*

CWW—BASIC COMPUTATIONAL PROCESS

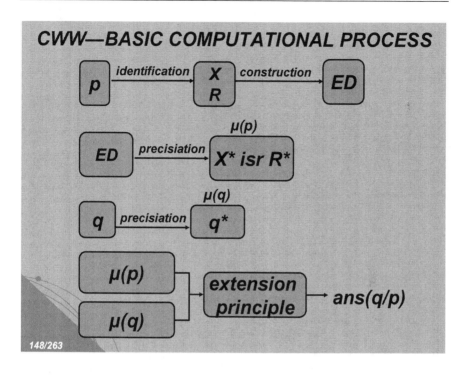

148/263

EXAMPLE. PROBLEM (a)

- **Note. In the following example restrictions are assumed to be possibilistic, r=blank.**

- **q: What is the average height of Swedes?**

  l=p: Most Swedes are tall

  Precisiation of p

  Clarification. Clarification not needed

149/263

EXAMPLE (DISCRETE VERSION)

Step 1. Identification (explicitation) of X and R.

X is identified as the proportion of tall Swedes among Swedes.

Correspondingly, R is identified as most.

- *To proceed further, it is necessary to digress to computation of the cardinality (count) of a fuzzy set.*

150/263

DIGRESSION

- *In fuzzy logic, proportion is defined as a relative ΣCount. (Zadeh 1983b) More specifically, if A and B are fuzzy sets in U, U={u₁, ..., uₙ}, the ΣCount(cardinality) of A is defined as:*

$$\Sigma Count(A) = \Sigma_i \mu_A(u_i)$$

151/263

The relative *ΣCount of B in A is defined as:*

$$\Sigma Count(B/A) = \frac{\Sigma Count(A \cap B)}{\Sigma Count(A)}$$

$$= \frac{\Sigma_i(\mu_A(u_i) \wedge \mu_B(u_i))}{\Sigma_i \mu_A(u_i)}$$

where $\cap$ =*intersection and* $\wedge$ =*min*
End of digression.

EXAMPLE (CONTINUED)

In application to the example under consideration, assume that the height of ith Swede, Name$_i$, is h_i and that the grade of membership of h_i in tall is $\mu_{tall}(h_i)$, i=1, ..., n. On this understanding, X may be expressed as:

$$X = \frac{1}{n}(\sum_{i=1}^{n} \mu_{tall}(h_i))$$

Step 2. Construction of ED.
The needed information is contained in the explanatory database, ED, where

ED= POPULATION.SWEDES[Name;
*   Height]+*
*   TALL[Height; μ]+*
*   MOST[Proportion; μ]*

Step 3. Precisiation of X and R.
In relation to ED, precisiated X may be
expressed as a function of ED

$$X^* = \frac{1}{n}\left(\Sigma_i\, \mu_{tall}(h_i)\right)$$

R* may be expressed as:

*   R* = most*

or equivalently, as:

*   R* = MOST[Proportion; μ]*

- **The precisiated canonical form is expressed as:**

CF*p: X* is R*

More specifically,

$$CF^*(p): \frac{1}{n}(\sum_{i=1}^{n} \mu_{tall}(h_i)) \quad \text{is most}$$

Equivalently,

$$\mu(p) = \mu_{most}(\frac{1}{n}(\sum_{i=1}^{n} \mu_{tall}(h_i)))$$

156/263

- **Precisiation of q**
q: What is the average height, h_{ave}, of Swedes?
q* may be expressed as a function of ED

$$q^*: ? h_{ave} = \frac{1}{n}\sum_{i=1}^{n} h_i = ans(q/I)$$

At this point, computation of ans(q/I) involves propagation of a possibilistic restriction

157/263

$$\frac{\frac{1}{n}(\sum_{i=1}^{n}\mu_{tall}(h_i))\quad \text{is most}}{\frac{1}{n}\sum_{i=1}^{n}(h_i)\quad \text{is ?}h_{ave}}$$

where h_{ave} is a fuzzy set.

- **Applying the extension principle, h_{ave} may be expressed as a solution of the variational problem:**

$$\mu_{h_{ave}}(v) = sup_h \, \mu_{most}\left(\frac{1}{n}\sum_{i=1}^{n}\mu_{tall}(h_i)\right)$$

subject to:

$$v = \frac{1}{n}\sum_{i=1}^{n}(h_i)$$

where $h = (h_1, ..., h_n)$

TRUTH-VALUE

The truth-value of p, t(p, ED+), is the degree to which the restriction X* isr R* is satisfied. More concretely,

$$t(p, ED^{+}) = \mu_{most}\left(\frac{1}{n}\Sigma_i\,\mu_{tall}(h_i)\right)$$

Note. The right-hand side of this equation may be viewed as a restriction on ED, with μ_{most}, μ_{tall} and h instantiated.

EXAMPLE. PROBLEM (a). CONTINUOUS VERSION

- In the discrete version of Problem (a), information about the height of Swedes is contained in the relation POPULATION.SWEDES[Name; Height]

- In the continuous version, this information is provided by the height density function, h, defined as:

  h(u)du=Proportion of tall Swedes whose height lies in the interval [u, u+du].

● *By definition,*

$$\int_{h\ min}^{h\ max} h(u)du = 1$$

● *In terms of h, precisiation of q may be expressed as:*

$$q^* : ? h_{ave} = \int_{h\ min}^{h\ max} uh(u)du$$

● *The explanatory database, ED, with suppressed arguments, may be expressed as:*

$$ED = h + \mu_{tall} + \mu_{most}$$

● *Correspondingly, the precisiated X, X*, and the precisiated R, R*, may be expressed as:*

$$X^* = \int_{h\ min}^{h\ max} \mu_{tall}(u)h(u)du$$

$$R^* = most$$

- **The precisiated canonical form reads:**

$$CF^*(p): \int_{h\,min}^{h\,max} \mu_{tall}(u)h(u)du \text{ is most}$$

- **Equivalently,**

$$\mu(p) = \mu_{most}\left(\int_{h\,min}^{h\,max} \mu_{tall}(u)h(u)du\right)$$

where $\mu(p)$ is the generalized intension of p.

- **At this point, we have**

$$\int_{h\,min}^{h\,max} \mu_{tall}(u)h(u)du \quad \text{is most}$$

$$\rule{6cm}{0.4pt}$$

$$\int_{h\,min}^{h\,max} uh(u)du \quad \text{is } ?h_{ave}$$

- **Applying the extension principle, we arrive at:**

$$\mu_{h_{ave}}(v) = sup_h\, \mu_{most}\left(\int_{h\,min}^{h\,max} \mu_{tall}(u)h(u)du\right)$$

subject to

$$v = \int_{h\ min}^{h\ max} uh(u)du$$

and

$$\int_{h\ min}^{h\ max} h(u)du = 1$$

- *Note that the continuous version is simpler to formulate than the discrete version.*

166/263

SUMMATION

- *In essence, Computing with Words is a system of computation in which the objects of computation are words, phrases and propositions drawn from a natural language.*

- *More concretely, Computing with Words adds to traditional systems of computation two important capabilities:*

167/263

a) *The capability to precisiate the meaning of words, phrases and propositions drawn from a natural language.*

b) *The capability to compute with precisiated words, phrases and propositions.*

- *Computing with Words has an important ramification for mathematics. It empowers mathematics with a new capability—a capability which traditional mathematics does not have—a capability to construct mathematical solutions of problems which are stated in a natural language.*

APPENDIX

EXAMPLE 1

Step 1. Precisiation of I (information set)

- *To begin with, let us rewrite I as:*

Prob(Height(John) is tall) is probable.

- *In this expression, Height(John) is a real-valued random variable, "Height(John) is tall" is an event described in a natural language, and probable is a probability—a probability which is described in a natural language.*

- *To proceed further it is necessary to precisiate tall and probable. The posed problem is trivialized if tall and probable are assumed to be interval-valued. Furthermore, assuming that tall and probable are interval-valued is a poor model of reality.*

- *It is much more realistic to assume that tall and probable are fuzzy sets with specified membership functions μ_{tall} and $\mu_{probable}$, respectively. It is the responsibility of whoever poses the problem to specify μ_{tall} and $\mu_{probable}$.*

● *What is the precisiated meaning of I?*
*Let p_H be the probability density function
of Height(John). In terms of p_H, the
probability measure of the fuzzy set tall
may be expressed as (Zadeh 1968):*

$$\int_R \mu_{tall}(u)p_H(u)du$$

*where R is the real line. Consequently, I
may be expressed as:*

$$\int_R \mu_{tall}(u)p_H(u)du \quad \text{is probable}$$

*with probable playing the role of a fuzzy
restriction on the probability measure of
tall. Equivalently, probable may be
interpreted as the possibility distribution
of the probability measure of tall. With
this interpretation of probable, the
restriction on p_H may be expressed as:*

$$\mu(p_H) = \mu_{probable}\left(\int_R \mu_{tall}(u)p_H(u)du\right)$$

where μ is the possibility distribution of p_H.

● *What can be concluded at this point is that the translation of I into a mathematical language may be expressed as:*

Probably John is tall $\xrightarrow{\text{precisiation}}$

$$\int_R \mu_{tall}(u)p_H(u)du \quad \text{is probable}$$

or equivalently as:

Probably John is tall $\xrightarrow{\text{precisiation}}$

$$\mu_{probable}\left(\int_R \mu_{tall}(u)p_H(u)du\right)$$

NOTE

- *Note that for given $\mu_{probable}$, μ_{tall} and p_H, the right-hand side of the above expression evaluates to a number in the interval [0, 1] which may be interpreted as the possibility of p_H or the truth-value of I given $\mu_{probable}$, μ_{tall} and p_H. It is important to observe that the expression:*

$$\mu(p_H) = \mu_{probable}\left(\int_R \mu_{tall}(u)p_H(u)du \right)$$

This expression plays the role of a semantic deep structure (generalized intension) of I (Zadeh 2006a). An abstracted version of the generalized intension—the protoform of I—reads:

$$\mu(p) = \mu_A\left(\int_R \mu_B(u)p(u)du \right)$$

NOTE

- *It is of interest to observe that the propositions "Probably John is tall," "Most Swedes are tall," and "Usually, it takes Robert about an hour to get home from work" have identical protoforms. What this observation brings to light is that seemingly very different propositions drawn from a natural language may have the same mathematical deep structure.*

Step 2. Precisiation of question(s)

- *We assume that short, very short and not very tall, are fuzzy sets with specified membership functions μ_s, μ_{vs} and μ_{nvt}, respectively. Using the technique employed for precisiation of I, precisiations of questions may be expressed as follows*

Version 1

Version 1.

- *What is the probability that John is short?* $\xrightarrow{precisiation}$

$$\int_R \mu_S(u)\,p_H(u)\,du \quad is \quad ?P_s$$

182/263

Version 2

Version 2.

- *What is the probability that John is very short?* $\xrightarrow{precisiation}$

$$\int_R \mu_{vs}(u)\,p_H(u)\,du \quad is \quad ?P_{vs}$$

183/263

Version 3

Version 3.

- **What is the probability that John is not very tall?** *precisiation*→

$$\int_R \mu_{nvt}(u)p_H(u)du \text{ is } ?P_{nvt}$$

where P_s, P_{vs} and P_{nvt}, are the probabilities that John is short, John is very short and John is not very tall, respectively.

Step 3. Computation of answer(s) to question(s).

- **In the case of Version 1 and Version 2, there is a shortcut. For these versions, it is expedient to compute the probability that John is not tall. The membership function of not tall, μ_{nt}, is related to that of tall, μ_t, by:**

$$\mu_{nt} = 1 - \mu_t$$

• *Consequently, the probability of the event "John is not tall," may be expressed as:*

Prob(Height(John) is not tall)

$$= \int_R \mu_{nt}(u)p_H(u)du$$

$$= \int_R (1 - \mu_t)p_H(u)du$$

186/263

$$= 1 - \int_R \mu_t(u)p_H(u)du$$

$$= 1 - probable$$

• *The fuzzy set short is a subset of not tall. Consequently, the probability measure of short is less than or equal to the probability measure of not tall. From this it follows that:*

187/263

$$= 1 - \int_R \mu_t(u) p_H(u)\,du$$

$$= 1 - probable$$

- The fuzzy set short is a subset of not tall. Consequently, the probability measure of short is less than or equal to the probability measure of not tall. From this it follows that:

$$P_s \ is \ \leq (1\text{-}probable)$$

188/263

- Note that 1-probable is a fuzzy set and that ≤ (1-probable) is the composition of ≤ and 1-probable. This is the answer to the question posed in Version 1.

- In Version 2, very short, like short, is a subset of not tall. Consequently, as in Version 1:

$$P_{vs} \ is \ \leq (1\text{-}probable)$$

189/263

- *It may appear to be counterintuitive that P_s and P_{vs} are identical. The reason for identity is that P_s and P_{vs} are not the values but fuzzy restrictions on the values of the probability measures of short and very short. In fact, the answer, $\leq$ (1-probable), is the same for all subsets of not tall.*

190/263

Version 3.

- *The shortcut does not work in the case of Version 3 since not very tall may not be a subset of not tall. What has to be employed in Version 3 is a version of the extension principle (Zadeh 1965, 1975a).*

- *The problem we are faced with may be expressed as an inference schema:*

191/263

$$\frac{\int_R \mu_t(u)p_H(u)du \quad \text{is probable}}{\int_R \mu_{nvt}(u)p_H(u)du \quad \text{is } ?P_{nvt}}$$

This schema is similar to the schema of the extension principle:

$$\frac{f(X) \text{ is } A}{g(X) \text{ is } ?B}$$

192/263

where f and g are given functions or functionals, and A and B are fuzzy sets. The extension principle asserts that the membership function of B may be expressed as a function of f, g and the membership function of A. More concretely,

$$\mu_B(v) = sup_u \, \mu_A(f(u))$$

subject to

$$v=g(u)$$

193/263

In application to Version 3, the extension principle leads to an answer to the question: What is the probability that John is not very tall? The answer is a fuzzy probability whose membership function may be expressed as:

$$\mu_{Pnvt}(v) = sup_{p_H} \mu_{probable}\left(\int_R \mu_t(u)p_H(u)du\right)$$

subject to

$$v = \int_R \mu_{nvt}(u)p_H(u)du$$

- *In conclusion, computation of the probability that John is not very tall reduces to the solution of a variational problem—a problem which involves maximization of a function of p_H subject to a side condition. What is important to note is that p_H is not a real-valued variable but a real-valued function.*

- *The solution of a discrete version of the posed problem reduces to the solution of a variational problem. The solution which is described is not at all in the spirit of standard probability theory. The posed problem is simple to state but not simple to solve.*

EXAMPLE 2 (Zadeh 2006a)

Swedes are much taller than Italians.

Clarification.

Most Swedes are much taller than most Italians.

Step 1. Identification of X and R.

X= Proportion of Swedes who are much taller than most Italians.

R= Most

Step 2. Construction of ED

POPULATION.SWEDES[Name; Height] +

POPULATION.ITALIANS[Name; Height] +

MUCH.TALLER[Height1; Height2; µ] +

MOST[Proportion; µ]

Note. In the following, $_AR$ denotes the projection of R and A. Equivalently, $_AR$ may be read as: Read A in R

Step 3. Precisiation of X and R

Precisiation of X

Find the height of ith Swede, NameS$_i$, i=1, ..., m

$h_i =_{Height}POPULATION.SWEDES[Name= NameS_i; Height]$

Find the height of jth Italian, NameI$_j$, j=1, ..., n

$k_j =_{Height}POPULATION.ITALIANS[Name= NameI_j; Height]$

EXAMPLE 3

Find the degree, a$_{ij}$, to which NameS$_i$ is much taller than NameI$_j$

$a_{ij} =_\mu MUCH.TALLER[Height_1=h_i; Height2=k_j; \mu]$

Compute the proportion, p$_i$, of Italians in the relation to whom NameS$_i$ is much taller

$$p_i = \frac{1}{n}\Sigma_j a_{ij}$$

Compute the degree, q_i, to which p_i satisfies most

$$q_i =_\mu MOST \, [Pr \, oportion = \frac{1}{n} \Sigma_j \, a_{ij}; \mu]$$

Compute the proportion, X^*, of Swedes who are much taller than most Italians

$$X^* = \frac{1}{m} \Sigma_i \, q_i$$

202/263

Precisiation of R

$R^* = MOST[Proportion; \mu]$

The precisiated canonical form may be expressed as $CF^*(p) = X^*$ is R^*, where

$$X^* = \frac{1}{m} \Sigma_i \, q_i$$

$$R^* = MOST[Proportion; \mu]$$

Additional examples may be found in cited papers.

203/263

INFORMAL EXPOSITION OF RS— CLARIFICATION DIALOGUE

- *The basic ideas which underlie precisiation of meaning and, more particularly, restriction-based semantics, are actually quite simple. To bring this out, it is expedient to supplement a formal exposition of RS with an informal narrative in the form of a dialogue between Robert and Lotfi. In large measure, the narrative is self-contained.*

DIALOGUE

Robert: Lotfi, restriction-based semantics looks complicated to me. Can you explain in simple terms the basic ideas which underlie RS?

Lotfi: I will be pleased to do so. Let us start with an example, p: Most Swedes are tall. p is a proposition. As a proposition, p is a carrier of information. Without loss of generality, we can assume that p is a carrier of information about a variable, X, which is implicit in p. If I asked you what is this variable, what would you say?

206/263

Robert: As I see it, p tells me something about the proportion of tall Swedes among Swedes.

Lotfi: Right. What does p tell you about the value of the variable?

Robert: To me, the value is not sharply defined. I would say it is fuzzy.

Lotfi: So what is it?

Robert: It is the word "most."

207/263

Lotfi: You are right. So what we see is that p may be interpreted as the assignment of a value "most" to the variable, X: Proportion of tall Swedes among Swedes.

Robert: I have a question. What is on your mind when you say that X is identified by inspection? Could you explain?

208/263

Lotfi: It is a good question. At this stage of the development of CWW, we do not have an algorithm for identifying X. For a human subject, identifying X is not a difficult task. If I am given a proposition, p, and am asked to identify X, I would ask myself the question: What is the variable which p constrains? Given my understanding of natural language and my world

209/263

knowledge, I could learn how to identify *X* after being shown a few examples, and so can everybody else.

Robert: Thank you for your explanation.

Lotfi: Your welcome. As you can see, a basic difference between a proposition drawn from a natural language and a proposition drawn from a mathematical language is that in the latter the variables and the values assigned to

them are explicit, whereas in the former the variables and the assigned values are implicit. There is an additional difference. When *p* is drawn from a natural language, the assigned value is not sharply defined—typically it is fuzzy, as "most" is. When *p* is drawn from a mathematical language, the assigned value is sharply defined.

Robert: I get the idea. So what comes next?

Lotfi: There is another important point. When p is drawn from a natural language, the value assigned to X is not really a value of X—it is a restriction (constraint) on the values which X is allowed to take. This suggests an unconventional definition of a proposition, p, drawn from a natural language. Specifically, a proposition is an implicit (hidden) restriction on an implicit (hidden) variable.

212/263

I should like to add that the restrictions which I have in mind are not standard constraints—they are so-called generalized constraints. Thus, a restriction may be equated to a generalized constraint. In the following, restrictions and generalized constraints are used interchangeably.

213/263

*Robert: What is a generalized constraint?
Why do we need generalized constraints?*
*Lotfi: A generalized constraint (restriction)
is expressed as:*

X isr R

*where X is the constrained (restricted)
variable, R is the constraining (restricting)
relation—typically a fuzzy set—and r is an
indexical variable which defines how R
restricts X. Let me explain why the
concept of a generalized constraint
(restriction) is needed in precisiation of
meaning of a proposition drawn from a
natural language.*

Standard constraints are hard in the sense that they have no elasticity. In a natural language, meaning can be stretched. What this implies is that to represent meaning, a constraint (restriction) must have elasticity. To deal with richness of meaning, elasticity is necessary but not sufficient. Consider the proposition: Usually most United flights leave on time.

216/263

What is the constrained variable and what is the constraining relation in this proposition? Actually, for most propositions drawn from a natural language a large repertoire of constraints is not necessary. What is sufficient are three so-called primary constraints and their combinations. The primary constraints are: possibilistic, probabilistic and veristic.

217/263

Here are simple examples of primary constraints:

- *Possibilistic constraint:*
Robert is possibly French and possibly German
- *Probabilistic constraint:*
With probability 0.75 Robert is German
With probability 0.25 Robert is French
- *Veristic constraint:*
Robert is three-quarters German and one-quarter French

The role of primary constraints is analogous to the role of primary colors: red, green and blue. In most cases, constraints are possibilistic. Possibilistic constraints are much easier to manipulate than probabilistic constraints.

Robert: Could you clarify what you have in mind when you talk about elasticity of meaning?

Lotfi: I admit that I did not say enough. Let me elaborate. In a natural language, meaning can be stretched. Consider a simple example, Robert is young. Assume that young is a fuzzy set and Robert is 30.

Furthermore, assume that in a particular context the grade of membership of 30 in young is 0.8. To apply young to Robert, the meaning of young must be stretched. To what degree? In fuzzy logic, the degree of stretch is equated to (1 - grade of membership of 30 in young.) Thus, the degree of stretch is 0.2.

Furthermore, the grade of membership of 30 in young is interpreted as the possibility that Robert is 30, given that Robert is young. What this implies is that the fuzzy set young defines the possibility distribution of the variable Age (Robert). Note that the fuzzy set young is a restriction on the values which the variable Age (Robert) can take.

It is in this sense that the proposition Robert is young is a possibilistic restriction on Age (Robert).

Now, in a natural language almost all words and phrases are labels of fuzzy sets. What this means is that in a natural language the meaning of words and phrases can be stretched, as in the Robert example.

It is in this sense that words and phrases in a natural language have elasticity. Another important point. What I have said so far explains why in the realm of natural languages most restrictions are possibilistic. This is equivalent to saying what I said already, namely, that in a natural language most words and phrases are labels of fuzzy sets.

Robert: Many thanks. You clarified what was not clear to me.

Lotfi: May I add that there is an analogy that may be of assistance. More specifically, the fuzzy set young may be represented as a chain linked to a spring, as shown in the next slide. The left end of the chain is fixed and the position of the right end of the spring represents the value of the variable Age (Robert).

226/263

The force that is applied to the right end of the spring is a measure of grade of membership. Initially, the length of the chain is 0, as is the length of the spring.

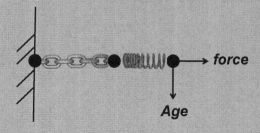

227/263

Robert: Many thanks for the explanation. The analogy helps to understand what you mean by elasticity of meaning.

Lotfi: I should like to add that elasticity of meaning is a basic characteristic of natural languages. Elasticity of meaning is a neglected issue in the literatures of linguistics, computational linguistics and philosophy of languages. There is a reason.

228/263

Traditional theories of natural language are based on bivalent logic. Bivalent logic, by itself or in combination with probability theory, is not the right tool for dealing with elasticity of meaning. What is needed for this purpose is fuzzy logic. In fuzzy logic everything is or is allowed to be a matter of degree.

229/263

Robert: Thanks again for the clarification. Going back to where we left of suppose I figured out what is the restricted variable, X, and the restricting relation, R. Is there something else that has to be done?

230/263

Lotfi: Yes, there is. You see, X and R are described in a natural language. What this means is that we are not through with precisiation of meaning of p. What remains to be done is precisiation (definition) of X and R. For this purpose, we construct a so-called explanatory database, ED, which consists of a collection of relations in terms of which X and R can be defined.

231/263

Robert: Can you be more specific?

Lotfi: To construct ED you ask yourself the question: What information—in the form of a collection or relations—is needed to precisiate (define) X and R? Looking at p, we see that to precisiate X we need two relations: POPULATION.SWEDES[Name; Height] and TALL[Height; μ].

In the relation TALL[Height; μ], μ is the grade of membership of a value of Height, h, in the fuzzy set tall. So far as R is concerned, the needed relation is MOST[Proportion; μ], where μ is the grade of membership of a value of Proportion in the fuzzy set most.

Equivalently, it is frequently helpful to ask the question: What is the information which is needed to assess the degree to which p is true?

234/263

At this point, we can express ED as the collection:

ED= POPULATION.SWEDES[Name; Height]+
    TALL[Height; μ]+
    MOST[Proportion; μ]

in which for convenience plus is used in place of comma.

235/263

Robert: So, we have constructed ED for the proposition, p: Most Swedes are tall. More generally, given a proposition, p, how difficult is it to construct ED for p?

Lotfi: For humans it is easy. A few examples suffice to learn how to construct ED. Construction of ED is easy for humans because humans have world knowledge. At this juncture, we do not have an algorithm for constructing ED.

236/263

Robert: Now that we have ED, what comes next?

Lotfi: We can use ED to precisiate (define) X and R. Let us start with X. In words, X is described as the proportion of tall Swedes among Swedes. Let us assume that in the relation POPULATION.SWEDES there are n names. Then the proportion of tall Swedes among Swedes would be the number of tall Swedes divided by n.

237/263

Here we come to a problem. Tall Swedes is a fuzzy subset of Swedes. The question is: What is the number of elements in a fuzzy set? In fuzzy logic, there are different ways of answering this question. The simplest is referred to as the ΣCount. More concretely, if A is a fuzzy set with a membership function μ_A, then the ΣCount of A is defined as the sum of grades of membership in A.

238/263

In application to the number of tall Swedes, the ΣCount of tall Swedes may be expressed as:

$$\Sigma Count(tall.Swedes) = \sum_{i=1}^{n} \mu_{tall}(h_i)$$

where h_i is the height of Name$_i$. Consequently, the proportion of tall Swedes among Swedes may be written as:

$$X = \frac{1}{n}(\sum_{i=1}^{n} \mu_{tall}(h_i))$$

239/263

This expression may be viewed as a precisiation (definition) of X in terms of ED. More specifically, X is expressed as a function of the variables h_1, ..., h_n, μ_{tall} and μ_{most}.

Precisiation (definition) of R is simpler. Specifically, R=most, where most is a fuzzy set. At this point, we have precisiated (defined) X and R in terms of ED.

240/263

Robert: So what have we accomplished?

Lotfi: We started with a proposition, p: Most Swedes are tall. We interpreted p as a possibilistic restriction (possibilistic constraint). We identified the restricted variable, X, as the proportion of tall Swedes among Swedes. We identified the restricting relation, R, as a fuzzy set, most. Next, we constructed an explanatory database, ED.

241/263

Finally, we precisiated (defined) X, R and q in terms of ED, that is, as a function of the variables $h_1, ..., h_n, \mu_{tall}$ and μ_{most}. In this way, we precisiated the meaning of p, which was our objective. The precisiated meaning may be expressed as the restriction:

$$\frac{1}{n}(\sum_{i=1}^{n}\mu_{tall}(h_i)) \text{ is most}$$

Robert: So, you precisiated the meaning of p. What purpose does it serve?

Lotfi: The principal purpose is the following. Unprecisiated (raw) propositions drawn from a natural language cannot be computed with. Precisiation is a prerequisite to computation. What is important to understand is that precisiation of meaning opens the door to computation with information described in a natural language.

Robert: Sounds great. I am impressed. However, it is not completely clear to me what you have in mind when you say "opens the door to computation with natural language." Can you clarify it?

Lotfi: With pleasure. Computation with natural language or, more or less equivalently, Computing with Words (CWW), is largely unrelated to natural language processing.

244/263

More specifically, computation with natural language is focused on computation with information described in a natural language. Typically, what is involved is solution of a problem which is stated in a natural language. Let me go back to our example, p: Most Swedes are tall. Given this information, how can you compute the average height of Swedes?

245/263

Robert: Frankly, your question makes no sense to me. Are you serious? How can you expect me to compute the average height of Swedes from the information that most Swedes are tall?

Lotfi: That is conventional wisdom. A mathematician would say that the problem is ill-posed and off limits to mathematics. It appears to be ill-posed for two reasons.

246/263

First, because the given information: Most Swedes are tall, is fuzzy, and second, because you assume that I am expecting you to come up with a crisp answer like "the average height of Swedes is 5' 10." Actually, what I expect is a fuzzy answer—it would be unreasonable to expect a crisp answer. The fuzzy answer is a restriction on the average height of Swedes, given that most Swedes are tall.

247/263

Equivalently, the fuzzy answer may be interpreted as the values of the average height of Swedes which are consistent with the information that Most Swedes are tall. As was stated earlier, in CWW consistency is equated to possibility, with the understanding that possibility is a matter of degree.

Robert: Thanks for the clarification. I am beginning to see the point of your question.

Lotfi: I should like to add a key point. The problem becomes well-posed if p is precisiated. This is the essence of Computing with Words.

Robert: I am beginning to understand the need for precisiation, but my understanding is not complete as yet. Can you explain how the average height of Swedes can be computed from precisiated p?

Lotfi: Recall that precisiated p is a possibilistic restriction expressed as:

$$\frac{1}{n}(\sum_{i=1}^{n}\mu_{tall}(h_i)) \text{ is most}$$

From the definition of a possibilistic restriction it follows that the restriction on X may be rewritten as:

$$t = \mu_{most}(\frac{1}{n}\sum_{i=1}^{n}\mu_{tall}(h_i))$$

What this expression means is that given the h_i, μ_{tall} and μ_{most}, we can compute the degree, t, to which the restriction is satisfied.

It is this degree, t, that is the truth-value of p. Now, here is a key idea. The precisiated p restricts X. X is a function of ED. It follows that ultimately what p restricts are variables in ED. This has important implications. Let me elaborate.

What we see is that the restriction induced by p on the h_i is of the general form

$f(h_1, ..., h_n)$ is most

What we are interested in is the induced restriction on the average height of Swedes. The average height of Swedes may be expressed as:

$$h_{ave} = \frac{1}{n}(\sum_{i=1}^{n} h_i)$$

This expression is of the general form

$$g(h_1, \ldots, h_n) \text{ is } ?h_{ave}$$

where $?h_{ave}$ is a fuzzy set that we want to compute.

At this stage, we can employ the extension principle of fuzzy logic to compute h_{ave}. (Zadeh 1975a) In general terms, this principle tells us that from a given possibilistic restriction of the form

$$f(x_1, \ldots, x_n) \text{ is } A$$

in which A is a fuzzy set, we can derive an induced possibilistic restriction on $g(x_1, \ldots, x_n)$,

$$g(x_1, \ldots, x_n) \text{ is } ?B,$$

in which B is a fuzzy set defined by the solution of the variational problem

$$\mu_B(v) = \sup_{x_1, \ldots, x_n} \mu_A(f(x_1, \ldots, x_n))$$

subject to

$$v = g(x_1, \ldots, x_n)$$

In application to our example, what we see is that we have reduced computation of the average height of Swedes to the solution of the variational problem

$$\mu_B(v) = \sup_{h_1, \ldots, h_n} \mu_{most}(f(h_1, \ldots, h_n))$$

subject to

$$v = \frac{1}{n}\left(\sum_{i=1}^{n} h_i\right)$$

In effect, this is the solution to the problem which I posed to you. As you can see, reduction of the original problem to the solution of a variational problem is not so simple.

However, solution of the variational problem to which the original problem is reduced, is well within the capabilities of desktop computers.

258/263

Robert: I am beginning to see the basic idea. Through precisiation, you have reduced the problem of computation with information described in a natural language—a seemingly ill-posed problem—to a well-posed tractable variational problem. I am impressed by what you have accomplished, though I must say that the reduction is nontrivial.

259/263

Without your explanation, it would be hard to see the basic ideas. I can also understand why computation with natural language is a move into a new and largely unexplored territory. Thank you for clarifying the import of your statement: precisiation of meaning opens the door to computation with natural language.

Lotfi: I appreciate your comment. May I add that I believe that in closed form the solution to the variational problem may be expressed as:

$$h_{max} \geq h_{ave} \geq most \times tall + (1\text{-}most)\, h_{min}$$

where most × tall is the product of fuzzy numbers most and tall, (1-most) h_{min} is the product of the fuzzy number (1-most) with the minimum height, h_{min}, and h_{max} is the maximum height.

Robert: This is a very interesting result, if true. It agrees with my intuition.

Lotfi: I appreciate your comment. I would like to conclude our dialogue with a prediction. As we move further into the age of machine intelligence and automated reasoning, the complex of problems related to computation with information described in a natural language, is certain to grow in visibility and importance.

262/263

The informal dialogue between Robert and Lotfi has come to an end.

263/263

References

Bardossy, A., Duckstein, L.: Fuzzy Rule-Based Modeling with Applications to Geophysical, Biological, and Engineering Systems. CRC Press, Boca Raton (1995)

Bargiela, A., Pedrycz, W.: Granular Computing. Kluwer Academic Publishers, Dordrecht (2002)

Chang, C.C., Keisler, H.: Model Theory. Studies in Logic and the Foundations of Mathematics, 3rd edn. Elsevier, Amsterdam (1990), 1973

Lambert, K., Van Fraassen, B.C.: Meaning relations, Possible Objects and Possible Worlds, Philos. Probl. Logic 1–19 (1970)

Lin, T.Y.: Granular Computing on Binary Relations II: Rough Set Representations and Belief Functions. In: Skowron, A., Polkowski, L. (eds.) Rough Sets In Knowledge Discovery, Physica-Verlag, pp. 121–140 (1998)

Mendel, J.: Uncertain Rule-Based Fuzzy Logic Systems: Introduction and New Directions. Prentice-Hall, Upper Saddle River (2001)

Takagi, T., Sugeno, M.: Fuzzy identification of systems and its application to modeling and control. IEEE Trans. Systems, Man, and Cybernetics 15, 116–132 (1985)

Yamakawa, T.: Stablization of an inverted pendulum by a high-speed fuzzy logic controller hardware system. Fuzzy Sets and Systems – On Applications of Fuzzy Logic Control to Industry 32(2) (1989)

Yen, J., Langari, R.: Intelligence, Control, and Information. Prentice-Hall, Englewood Cliffs (1999)

Zadeh, L.: Fuzzy sets. Information and Control 8, 338–353 (1965)

Zadeh, L.: Outline of a new approach to the analysis of complex systems and decision processes. IEEE Trans. on Systems, Man and Cybernetics SMC 3, 28–44 (1973)

Zadeh, L.: The concept of a linguistic variable and its application to approximate reasoning, Part I: Information Sciences 8, 199-249, 1975; Part II: Information Sciences 8, 3010-357, 1975; Part III: Information Sciences 9, 43–80 (1975a)

Zadeh, L.: Calculus of fuzzy restrictions, Fuzzy sets and Their Applications to Cognitive and Decision Processes. In: Zadeh, L.A., Fu, K.S., Tanaka, K., Shimura, M. (eds.) New York: Academic Press, pp. 1–39. Academic Press, London (1975b)

Zadeh, L.: Fuzzy sets as a basis for a theory of possibility. Fuzzy Sets and Systems 1, 3–28 (1978)

Zadeh, L.: Test-score semantics for natural languages and meaning representation via PRUF. In: Rieger, B. (ed.) Empirical Semantics, pp. 281–349. Brockmeyer, Bochum (1981). Also Technical Memorandum 246, AI Center, SRI International, Menlo Park, CA, 1981.

Zadeh, L.: A fuzzy-set-theoretic approach to the compositionality of meaning: propositions, dispositions and canonical forms. Journal of Semantics 3, 253–272 (1983a)

Zadeh, L.: A computational approach to fuzzy quantifiers in natural languages. Computers and Mathematics 9, 149–184 (1983b)

Zadeh, L.: Fuzzy logic = computing with words. IEEE Transactions on Fuzzy Systems 2, 103–111 (1996)

Zadeh, L.: Toward a theory of fuzzy information granulation and its centrality in human reasoning and fuzzy logic. Fuzzy Sets and Systems 90, 111–127 (1997)

Zadeh, L.: Some reflections on soft computing, granular computing and their roles in the conception, design and utilization of information/intelligent systems. Soft Computing 2, 23–25 (1998)

Zadeh, L.: From computing with numbers to computing with words—from manipulation of measurements to manipulation of perceptions. IEEE Transactions on Circuits and Systems 45, 105–119 (1999)

Zadeh, L.: A new direction in AI—toward a computational theory of perceptions. AI Magazine 22(1), 73–84 (2001)

Zadeh, L.: Toward a perception-based theory of probabilistic reasoning with imprecise probabilities. Journal of Statistical Planning and Inference, Elsevier Science 105, 233–264 (2002)

Zadeh, L.: Precisiated natural language (PNL). AI Magazine 25(3), 74–91 (2004)

Zadeh, L.: Generalized theory of uncertainty (GTU)—principal concepts and ideas. Computational Statistics and Data Analysis 51, 15–46 (2006a)

Zadeh, L.: From search engines to question answering systems—The problems of world knowledge, relevance, deduction and precisiation. In: Sanchez, E. (ed.) Fuzzy Logic and the Semantic Web, Ch. 9, pp. 163–210. Elsevier, Amsterdam (2006b)

Zadeh, L.: A note on Z-numbers. Information Sciences 181, 2923–2932 (2011)